B

AVON COUNTY LIBRARY
WITHDRAWN AND OFFERED FOR SALE
SOLD AS SEEN

-4 NOV 1985

STEAMING INTO THE EIGHTIES

The Standard Gauge Railway Preservation Scene

G T Heavyside

DAVID & CHARLES
Newton Abbot   London   North Pomfret (Vt)   Vancouver

B 033880

British Library Cataloguing in Publication Data

Heavyside, G T
  Steaming into the eighties.
  1. Locomotives—Great Britain—History
  I. Title
  625.2'61   TJ603.4.G7

ISBN 0-7153-7513-X

Library of Congress Catalog Card Number: 78-62482

© G T Heavyside 1978

All rights reserved. No part of this publication may be reproduced, stored in a retrieval system, or transmitted, in any form or by any means, electronic, mechanical, photocopying, recording or otherwise, without the prior permission of David & Charles (Publishers) Limited

Printed in Great Britain
by Biddles Limited, Guildford, Surrey
for David & Charles (Publishers) Limited
Brunel House   Newton Abbot   Devon

Published in the United States of America
by David & Charles Inc
North Pomfret   Vermont 05053   USA

Published in Canada
by Douglas David & Charles Limited
1875 Welch Street   North Vancouver   BC

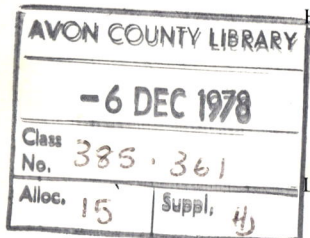

Contents

Introduction   5

Preserved Lines and Museums   6

Location map, Strathspey Railway, Scottish Railway Preservation Society, North of England Open Air Museum, Tanfield Railway, North Yorkshire Moors Railway, National Railway Museum, Derwent Valley Railway, Yorkshire Dales Railway, Keighley & Worth Valley Railway, Midland Railway Trust, Main Line Steam Trust, Shackerstone Railway Society, Nene Valley Railway, North Norfolk Railway, Bressingham Steam Museum, Stour Valley Railway Preservation Society, Kent & East Sussex Railway, Bluebell Railway, Great Western Society, Isle of Wight Steam Railway, Winchester & Alton Railway, Day to Day Running, Dart Valley Railway, West Somerset Railway, East Somerset Railway, Bristol Suburban Railway, Dean Forest Railway Society, Planning the Future, Severn Valley Railway, Chasewater Light Railway, Foxfield Light Railway, Bury Transport Museum, Steamtown, Lakeside & Haverthwaite Railway.

Return to the Main Line   60

Map BR Steam Special Routes 1972—1977, Scotland, Newcastle—Middlesbrough, Newcastle—Carlisle, York—Scarborough, Scarborough—Hull, Leeds—York, Organising a Steam Special, Manningtree—March, Westbury—Eastleigh, Salisbury—Basingstoke, Birmingham—Didcot, Birmingham—Stratford-upon-Avon, Oxford—Hereford, Newport—Chester, Sheffield—Guide Bridge and Dinting, Leeds—Carnforth, Carnforth—Sellafield.

Ireland   90

1975—Rail 150 Celebrations   92

The Might and Majesty of Steam   96

Introduction

'Steam for ever', 'long live steam', were slogans frequently chalked on smoke-box doors of locomotives eking out their last days on British Rail during the 1960s. Their passing from BR was genuinely mourned, and those messages written with sincerity and affection, but the scribes could hardly have visualised what great words of prophesy they wrote.

The steam locomotive, often described as the nearest mechanical creation to an animal, had given the country's railways nearly 150 years of faithful service, and endeared itself to the hearts of many people. With its impending demise from BR (except on the narrow gauge Vale of Rheidol line) various people were loath to see it become just another page in the history books, or confined to static exhibition, as had usually been the case with the early preserved examples, and thus was born, at the turn of the decade into the sixties, the standard gauge preservation movement as we know it today.

Progress in this new epoch has been phenomenal, and is all the more remarkable in that it has mostly been accomplished through kind benefactors and volunteer labour, whose only dividend is the joy of working with steam and the satisfaction of knowing that their efforts are helping to ensure its future. Today over 550 standard gauge locomotives, which are steamable or potential workers (although many still require extensive renovation) can be found at centres throughout the UK, the geographical distribution of the operational lines and museums being seen from a glance at the map on page 12. Total assets run into several million pounds and some of the major tourist railways have passenger journey totals of more than ¼ million each annually, which means that two to three million people travel behind steam purely for enjoyment on both standard and narrow gauge lines every year. At the moment the growth shows little sign of abatement, with more engines being preserved, further centres projected, and many plans for extensions.

The revival of steam railways has not been without problems, and in some instances, as in the nineteenth century, today's restorers have met with hostility from many quarters, but once established they have become an integral part of an area's amenities, widely publicised by the tourist authorities, both in Britain and abroad. There is much co-operation with BR and some engines are allowed to haul trains over selected main lines.

In this book I have endeavoured to trace the history of the standard gauge preservation movement and to depict steam in its new environment, both on private metals, and on some of its many excursions over BR tracks. We take a brief look at the Rail 150 celebrations of 1975 and the Irish scene.

Steam centres today form part of the widespread activity to conserve many aspects of our inheritance, but the book has been compiled with no sense of nostalgia, as Trevithick's brainchild is still a live force, and from the preservation viewpoint, although only time will confirm, probably only just emerging from adolescence. In reviewing and lauding the efforts in this field to date, let us at the same time wish the movement *bon voyage*, not only into the eighties, but *ad infinitum*.

I would like to place on record my sincere appreciation to the many people—far too numerous to mention by name—who have willingly given assistance towards the compilation of this book, including the photographers who loaned prints. The publication of some of these photographs has enabled me to depict a far broader picture than would otherwise have been possible.

I am very much indebted to Messrs G M Kichenside, C B Phillips and D C Williams, for giving an insight into aspects of preservation which are not often fully appreciated, but so vitally necessary.

My special thanks are due to Mr Chris Makin and Mrs Mavis Phillips for their work on the prints and Mrs Marlene McPherson for kindly typing the manuscript.

Bolton, March 1978　　　　　　　　G T Heavyside

Kent & East Sussex Railway 0-6-0ST No 23 waits to leave Rolvenden with a train for Tenterden on 15 September 1974. The engine was built by Hunslet No 3791 in 1952, and over 30 of the 484 engines constructed to this highly successful 1942 austerity design can be found at preservation sites throughout the UK, while others still work daily at various collieries. The KESR was the first line built under the 1896 Light Railways Act, the inaugural section being opened in 1900, and in honour of the line's engineer and first general manager No 23 has since been named Holman F Stephens.　　(G T Heavyside)

Preserved Lines and Museums

When the feature film *The Titfield Thunderbolt* was released in 1953 it must have seemed that only a group of eccentrics could contemplate such an outlandish scheme as that portrayed, that is reopening by private means a forsaken British Railways branch line. The story of how the Vicar of Titfield and his local followers were successful in keeping trains running to Mallingford, and revived an ancient engine from a nearby museum is simply film legend, but in reality the theme has since been re-enacted many times. By 1953 private preservation of the narrow gauge Talyllyn Railway was well established, soon to be followed by the Festiniog, but standard gauge by comparison was a totally different proposition. However as history has frequently proved, the improbable pipe-dreams of today often become the everyday occurrences of tomorrow.

In August 1960 the first regular steam passenger service on a preserved standard gauge line began, when the Bluebell Railway, run entirely by volunteers, commenced operations from Sheffield Park to Bluebell Halt, near Horsted Keynes station, Sussex, the latter station then still being used by BR. By today's standards, traffic was restored in an incredibly short time, the inaugural meeting of the society only taking place in March 1959. Two months before instigation of Bluebell Railway trains a new management became responsible for trains over the historic Middleton Railway in Leeds, although the predominant purpose was the carriage of freight, generally by diesel locomotives, between BR and local industrial concerns, work which continues on a limited scale today.

The first fruits of the seeds sown in the late 1950s had matured, but the movement was only in its infancy. Steam was very much alive on BR and its total withdrawal as forecast in the 1955 Modernisation Plan seemed somewhat remote, but a few far-sighted individuals had by this time launched the Railway Preservation Society, with the aim of fostering and co-ordinating development of standard gauge sites throughout the UK, this subsequently becoming the responsibility of the Association of Railway Preservation Societies.

The announcement in late 1960 of a further 27 steam engines to be officially preserved by the then British Transport Commission, supplementing the 44 already in its care, shook many from their complacency. The list was controversial regarding both inclusions and omissions, and it was obvious that if representatives of many well loved classes were to be saved, then it would have to be by private enterprise.

Following the example of Capt W G Smith who purchased former Great Northern Class J52 0-6-0ST No 68846 in 1959 a few engines were bought by individuals, including the renowned A3 Pacific No 60103 *Flying Scotsman* by Alan Pegler in 1963. The high cost involved, often running into thousands, was prohibitive to most people, but various societies were formed with the aim of raising the necessary funds to compete with the scrap metal merchants. Time was usually short if BR deadlines were to be met, but with the majority of supporters anxious to see engines restored to working order there was every incentive to support the numerous appeals, although in order to achieve the initial objectives many large loans had to be procured. There was much frantic activity but by the time BR finally dispensed with steam in August 1968 many otherwise doomed locomotives had been promised a new lease of life.

Meanwhile BR had also closed many unremunerative lines, and while there was no shortage of proposals for reopening some of these lines, only a further three ex BR branches saw regular operation again by the end of the decade. These were the Lochty Railway, Fifeshire, inaugurated in June 1967, followed by the Worth Valley from Keighley in June 1968, after six years of negotiations, and, in the following April, the Dart Valley Railway, which commenced services from Buckfastleigh towards Totnes with the aim of keeping alive a typical GWR branch line. The three provide stark contrasts in their constitutions, the Lochty being a private railway, the KWVR virtually operated by volunteers, while the DVR is run as a commercial venture, although still relying on a certain amount of voluntary labour.

Since then the pessimists have regularly cried 'enough' and that consolidation should be the keynote, but the pleas have largely gone unheeded, with more schemes being projected. Indeed future historians may well look back on the mid-1960s and 1970s as the second railway mania, and with an examination of plans that

The Bluebell Railway took delivery of its first stock on 17 May 1960 when 'Terrier' Class A1X 0-6-0T No 55 and two coaches (an LSWR third and an SR corridor brake) arrived, here seen approaching Horsted Keynes along the electrified branch from Haywards Heath. The Bluebell lost its connection with BR when this line was closed on 28 October 1963. *(Derek Cross)*

would have made even George Hudson shudder, such as the abortive attempt to reopen the Waverley route. There is a long list of abandoned ideas, but about 20 operating lines have, nevertheless, successfully emerged from the embryo stage, together with numerous other centres which usually offer short rides behind steam. Working steam also features at museums which house various modes of transport, and others where it is part of a broader plan to portray a wide spectrum of life from the past.

Development has not been without tribulation, for the scene greeting the modern navvies has frequently been one of almost total desolation, few if any facilities, and track lost in a jungle of weeds. Hard work has been the essence, if permanent way, buildings, locomotives and stock, etc, were to be restored to an acceptable and safe standard for a discerning public, and satisfy the rigorous and necessary requirements of the Railway Inspectorate to grant running rights. Legal requirements must be observed and authority to operate is generally obtained under the Light Railways Acts of 1896 and 1912, the initial application usually being made by BR with a subsequent transfer of the order to the new company.

However, before the promoters have obtained the requisite powers they have often faced many objections, from individuals who simply feared the return of the iron horse, to statutory bodies intent on using the land as car parks, footpaths, etc. These disagreements have in some cases culminated in public enquiries and in others litigation (the Kent & East Sussex Railway being involved in a legal battle with the Ministry of Transport for over three years), which has created much additional, and oft-times unnecessary expense, simultaneously delaying opening day and depriving the railways of revenue. Within a few of the new railways there have been battles for control by opposing factions, each passionately convinced of its own case. Today, as railways have grown up, so they have matured.

The continued expansion of the standard gauge preservation movement has only been possible because of the availability of further suitable steam locomotives and it may surprise

some that the supply of ex BR engines did not by any means become exhausted in 1968. The policy of the Barry scrap-metal merchants Woodham Bros is the principal reason for this, for while their contemporaries quickly reduced their consignments of redundant BR locomotives to fragments, the majority of those arriving at Barry, after removal of any high value metals, have stood broadly intact, braving the salty sea air, some for over 15 years. Thousands have made pilgrimages to South Wales to stand and stare, and dream of past glories, and so too have men with vision, and plenty of courage, for since September 1968 nearly 90 (about 33 per cent of the preserved BR locomotives) have departed for new homes, ranging from small tanks to BR 8P Pacific No 71000 *Duke of Gloucester*. About 130 engines remain at what has perhaps become the world's most famous scrap-yard, some the subject of further appeals.

The preservationists have also scoured the industrial lines where the changeover to diesel traction has been less traumatic than on BR and where small pockets of steam survive today. Over 300 engines, including a number of former BR examples, have been obtained from this source, and while lacking the romance of their main line sisters, many have proved themselves as suitable work horses in their new surroundings. These engines are generally of small stature and thus less costly to purchase, or maintain, and they monopolise sites such as Foxfield, Shackerstone and Tanfield. Some firms have donated or loaned their engines to the societies.

In their search for suitable power the adventurous have not been content to stay at home and 14 engines have been imported from Europe, including two from behind the Iron Curtain. Continental railways usually built their locomotives to the larger Berne loading gauge (roundly 1ft wider and higher than the British loading gauge) and to overcome operational difficulties the Nene Valley Railway, near Peterborough, has adopted this standard by making one or two structural alterations to platform faces or bridges. Five engines from overseas can be found here.

It is appropriate to mention the National Railway Museum, York, an outstation of the Science Museum under the auspices of the Department of Education & Science, since it plays an important role in the preservation movement. When the decision to close Clapham Museum of Transport and move the locomotives to York was mooted in a White Paper of November 1967 it engendered much controversy, many feeling that at least part of the official

Activity on another of the pioneers. NER 0-4-0T No 1310, of 1891 vintage, propels passengers travelling in freight stock from Tunstall Road Halt to Middleton Park Gates, on the Middleton Railway, Leeds, on 25 October 1975. In 1758 this was the first railway to be sanctioned by Act of Parliament and in 1812 the first to use steam locomotives regularly, albeit on the rack-rail principle. When the M1 motorway was extended into Leeds, a tunnel was built to accommodate the line under the road. *(G T Heavyside)*

collection should remain in London, and naturally other towns put forward their claims to have the museum. However the move has without doubt been a huge success, for within a year of opening in September 1975 the two millionth visitor had been welcomed. Its influence has been felt far beyond York, for engines have been loaned to various lines for operating purposes, and some have reciprocated, enabling the museum on occasions, to display types not in its own collection.

On view at York is Coronation Pacific No 6229 *Duchess of Hamilton*, previously exhibited at Butlins holiday camp, Minehead, before being loaned to the NRM. Three more large LMS engines and four SR tanks were additional static attractions at four camps and the decision to disperse them, during the 1970s, to operating sites (some have already come to life) provided another welcome source of motive power.

With resources often sparse, restoring the majority of engines has been a long, arduous, spare-time task, sometimes spread over many years, although a handful have received attention in BREL works and from other firms. Many laymen have become adept at locomotive renovation and maintenance, and indeed at various other railway operating skills, but the finished product has not always pleased the purists, with some engines carrying liveries foreign to their backgrounds, and a few bedecked with original colour schemes. However historical considerations are sometimes of secondary importance to keeping the stud in a reliable state to work traffic, and with this in mind some engines have had modifications carried out, including conversions to oil burning.

If a railway is to function successfully, or even if a small museum is to operate trains, then as well as engines and carriages, property, signalling, track-bed, lineside structures, vegetation, and sundry other items, must not be neglected, and to ensure this the larger concerns have formed departments to cover numerous operating activities. The fixed assets at the outset were usually minimal, which has meant that signals, signal boxes, water towers, turntables, even station buildings, etc, have had to be salvaged elsewhere, and sometimes transported hundreds of miles to their new quarters, while other structures have been purpose built to serve the new era, including numerous sheds for the stock.

It will readily be realised that the total cost of establishing even a small depot are astronomical and with galloping inflation and contract dates to meet, some have had to be content, at least initially, with a more modest scheme than originally envisaged. The Winchester & Alton Railway suffered this experience when in May 1975 it floated a public company aimed at raising a minimum of £625,000 but which failed to gain the necessary support, although a second issue in November 1975 raised over £75,000 allowing the purchase of three miles of track and just the track-bed of an adjacent seven miles for possible future extensions, a section it had hoped to operate from an early date. With such large sums involved and private railway operation growing into what has become big business, financial protection of members and long term security of assets is essential, and with this in

This group of discarded BR engines slowly decaying at Woodham Bros scrapyard, Barry, present a sorry spectacle on 24 March 1976, but hopefully, phoenix like, some will one day steam again. Left to right Nos 7200, 4270, 30506, 42765, 7927 and 34081. Certainly many of today's working engines looked like this for a time.
(G T Heavyside)

mind many limited companies and trusts have been formed. The advice of legal and accountancy professions has become indispensable.

To further development every avenue of finance and assistance has had to be explored, with grants being obtained from various bodies, including tourist boards, while industry has sponsored many projects from the movement and restoration of engines to the founding of full scale operating depots, as at H P Bulmer Ltd, the Hereford cider makers. The West Somerset Railway has benefited through Somerset County Council purchasing the land and permanent way and subsequently leasing this to the railway, while under the Government Manpower Services Commission Job Creation Scheme many centres have received grants (already in total over £500,000 and more in the pipeline) permitting completion of tasks which otherwise might have taken years. Army units have completed a number of projects as training exercises and similarly apprentices from a Peterborough engineering firm have restored Nene Valley Railway locomotives.

Extra revenue, together with valuable publicity, has come from the film industry, and the number of feature films, television serials, documentaries and commercials made on the steam lines are legion. Locomotives, stations, etc, have often appeared in unfamiliar roles, disguised on occasions as if they were part of the Victorian era, or as overseas railways, while LMS Class 5 4-6-0 No 45212 on the Keighley & Worth Valley Railway once starred in a television commercial adorned with wallpaper.

Welcome as these sources of income are, for their bread and butter and long term viability, the railways depend on attracting sufficient travellers on a regular basis, and these are people mainly intent on enjoying a day's recreation; none of the plans for commuter services having as yet materialised, apart from a short lived venture on the Torbay & Dartmouth Railway when the line was first acquired. With

Llangollen station with stock of the Llangollen Railway Society, which hopes gradually to introduce passenger services towards Berwyn and beyond. *(G M Kichenside)*

the sophisticated needs of the modern tourist demanding something more than just a basic train ride, encouragement is given to view the stored engines and those undergoing restoration, small displays of railwayana have been established, along with souvenir shops, restaurants, and picnic areas, which also help to stimulate and sustain interest.

The lines and centres also stage additional attractions from time to time, including displays of other forms of vintage transport, schools specials, evening wine-and-dine trains, charters for wedding receptions and anniversaries, while at Christmas Santa Claus is a familiar visitor. Popular regular events are enthusiasts' days with more than the usual norm of engines in steam, 12 at one Severn Valley Railway occasion, and sometimes featuring demonstration freight trains, which apart from track maintenance trains, etc, are a rarity, except on the aforementioned Middleton Railway, and the Nene Valley Railway, which conveys a weekly consignment of insulation material to a local warehouse on the final stage of its journey from Stirling.

That the steam railways have successfully adapted to their new place in society can be gauged by the need to steam up to 60 engines at peak periods such as bank holiday week-ends, while passenger statistics on some lines annually break the six figure barrier. Patronage is such in some instances, that more intensive services are operated than ever before, and with the continuing interest of steam enthusiasts both old and new and the general public, augurs well for the future.

In addition these private lines stand today as living monuments, not only to Trevithick, Hackworth, the Stephensons, and the other early pioneers, but also to the likes of Stanier, Bulleid, Gresley, Riddles, and other latter day locomotive engineers, and through the spirit of this new regime the career of steam is still unfolding. May the last rites never be pronounced.

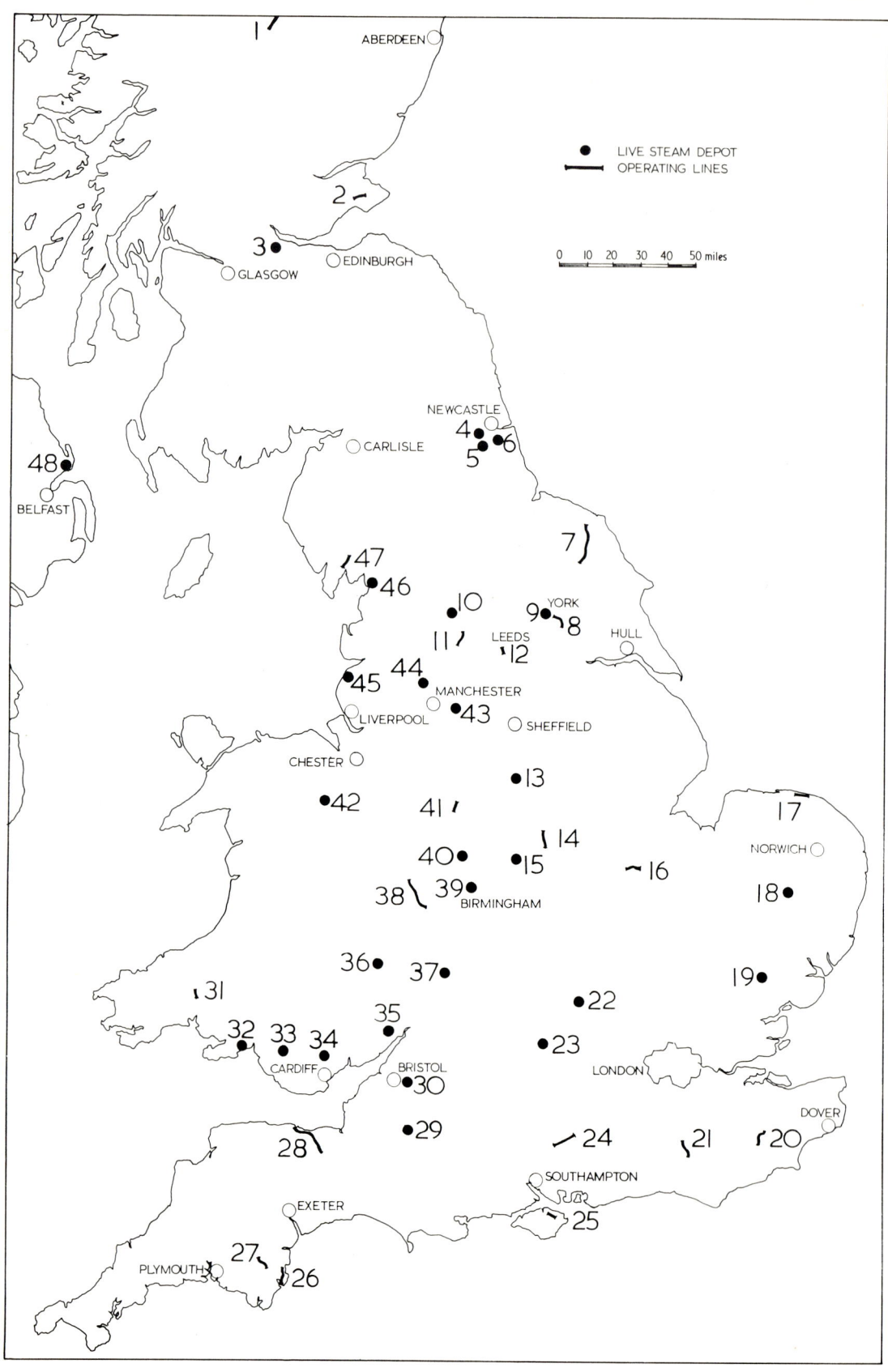

KEY TO MAP

1. Strathspey Railway. Aviemore—Boat of Garten.
2. Lochty Private Railway, Fifeshire. Lochty—Knightsward.
3. Scottish Railway Preservation Society, Falkirk.
4. Tanfield Railway, Marley Hill, Tyne & Wear.
5. North of England Open Air Museum, Beamish, Co Durham.
6. Bowes Railway, Springwell, Tyne & Wear.
7. North Yorkshire Moors Railway. Grosmont—Pickering.
8. Derwent Valley Railway. York Layerthorpe—Dunnington.
9. National Railway Museum, York.
10. Yorkshire Dales Railway, Embsay, Skipton.
11. Keighley & Worth Valley Railway. Keighley—Oxenhope.
12. Middleton Railway, Leeds. Tunstall Road Halt—Middleton Park Gates.
13. Midland Railway Trust, Butterley, Ripley, Derbyshire.
14. Main Line Steam Trust. Loughborough—Rothley.
15. Shackerstone Railway Society, Shackerstone, Leicestershire.
16. Nene Valley Railway. Wansford—Orton Mere, Peterborough.
17. North Norfolk Railway. Sheringham—Weybourne.
18. Bressingham Steam Museum, Diss, Norfolk.
19. Stour Valley Railway Preservation Society, Chappel & Wakes Colne, Essex.
20. Kent & East Sussex Railway. Tenterden—Wittersham Road.
21. Bluebell Railway, Sussex. Sheffield Park—Horsted Keynes.
22. Quainton Railway Society, Quainton, Bucks.
23. Great Western Society, Didcot, Oxfordshire.
24. Winchester & Alton Railway. Alresford—Ropley.
25. Isle of Wight Steam Railway. Havenstreet—Wootton.
26. Dart Valley Railway—Torbay & Dartmouth Line. Paignton—Kingswear.
27. Dart Valley Railway—Buckfastleigh Line. Buckfastleigh—Totnes (no access at Totnes).
28. West Somerset Railway. Minehead—Bishops Lydeard.
29. East Somerset Railway, Cranmore.
30. Bristol Suburban Railway, Bitton.
31. Gwili Railway, Carmarthenshire. Bronwydd Arms—Cwmdwyfran.
32. Swansea Maritime & Industrial Museum, Swansea.
33. 9642 Preservation Group, Maesteg, Mid Glamorgan.
34. Caerphilly Railway Society, Caerphilly, Mid Glamorgan.
35. Dean Forest Railway Society, Parkend and Norchard (two sites), Gloucestershire.
36. Bulmer Railway Centre, Hereford.
37. Dowty Railway Preservation Society, Ashchurch, Gloucestershire.
38. Severn Valley Railway. Bridgnorth—Foley Park, Kidderminster.
39. Birmingham Railway Museum, Tyseley, Birmingham.
40. Chacewater Light Railway, Brownhills, Staffordshire.
41. Foxfield Light Railway, Staffordshire. Dilhorne—Blythe Bridge.
42. Llangollen Railway Society, Llangollen, Clwyd.
43. Dinting Railway Centre, Dinting, Glossop, Derbyshire.
44. Bury Transport Museum, Bury, Greater Manchester.
45. Steamport, Southport, Merseyside.
46. Steamtown, Carnforth, Lancashire.
47. Lakeside & Haverthwaite Railway. Lakeside—Haverthwaite, Cumbria.
48. Railway Preservation Society of Ireland, Whitehead, Co Antrim.

Above: The Strathspey Railway hopes shortly to reinstate regular services from Aviemore to Boat of Garten on the original Highland Railway main line to Inverness. Here LMS Black 5 4-6-0 No 5025, representative of a class that did yeoman service on the line for many years, shunts empty stock at Boat of Garten on 29 May 1976. *(Bill Roberton)*

Left: Restored to its Caledonian Railway splendour McIntosh 0-4-4T No 419 bustles about the Scottish Railway Preservation Society depot at Falkirk on 29 August 1976. *(G T Heavyside)*

Above: At Beamish, County Durham, the North of England Open Air Museum portrays many aspects of life in the North East during the latter part of the nineteenth century. Railways feature prominently, and on 4 September 1977 NER class C 0-6-0 No 876, with period stock, storms past Rowley station, the buildings having been brought brick by brick from their original site just west of Consett. In the colliery area the replica *Locomotion*, built from drawings of the original for the Rail 150 celebrations, works on another section of track, although occasionally it is loaned to other sites for short periods. *(G T Heavyside)*

Right: On the same day, at the nearby Tanfield Railway, 1937 built Hudswell Clarke 0-4-0ST No 1672 *Irwell* passes Marley Hill Colliery with a coach constructed on site. *Irwell* was formerly in CEGB service and is one of a galaxy of ex-industrial engines at this location.
(G T Heavyside)

Above: The 18 mile North Yorkshire Moors Railway from Grosmont to Pickering, one of the most scenic in England and contained almost wholly within the North Yorkshire Moors National Park, had a royal official opening by the Duchess of Kent on 1 May 1973. On 8 August 1975 the first engine to be purchased privately from BR, Ivatt GNR Class J13 (LNER J52) 0-6-0ST No 1247 makes steady progress at Thomason Foss with a Grosmont—Goathland train. *(R E B Siviter)*

Left: Class K1 Mogul No 2005 in LNER apple green livery, emerges from Grosmont Tunnel, and passes No 1247 on 13 June 1976. *(G T Heavyside)*

Named after the eminent railway photographer *Eric Treacy*, LMS Black 5 4-6-0 No 5428 prepares to back down to Grosmont station on 3 August 1975. On the right outside the NYMR-erected shed are former Lambton Railway 0-6-2Ts Nos 5 and 29.
(G T Heavyside)

NER Class P3 0-6-0 No 2392 rounds the curve at Darnholm ready to make the final assault on the 1 in 49 climb to Goathland from Grosmont on Enthusiasts' Day, 1 May 1977. *(G T Heavyside)*

In desolate surroundings NER T2 0-8-0 No 2238 approaches the summit at Eller Beck, working through to Pickering on 1 May 1977. *(G T Heavyside)*

On the same day, with the 11.35 Pickering—Grosmont, the solitary Black 5 fitted with outside Stephenson link motion No 4767 *George Stephenson* leaves Levisham, in the magnificent glacial gorge of Newton Dale.
(G T Heavyside)

The stalls round a turntable of the former York North motive power depot provides an admirable location for the historic collection of locomotives at the National Railway Museum. On the right of this view is the world steam speed record holder, LNER No 4468 *Mallard*, which attained 126mph on 3 July 1938. *(G T Heavyside)*

During 1977 regular passenger services returned to the independent Derwent Valley Railway (which escaped nationalisation) after an absence of 51 years. In a sylvan setting at Murton on 21 May 1977 Class J72 0-6-0T No 69023 *Joem*, in NER livery, trundles by with the daily York Layerthorpe to Dunnington train. *(G T Heavyside)*

With steam to spare 1952-built Hunslet 0-6-0ST No 3715 *Primrose No 2* passes Embsay station, Skipton, on the Yorkshire Dales Railway, while Hudswell Clarke 0-6-0ST No 1709 of 1939 *Slough Estates Ltd No 5* (nearest camera) brings up the rear on 25 August 1975. On the left is Hudswell Clarke 0-6-0T No 1822 of 1949.
(G T Heavyside)

The Keighley & Worth Valley Railway opened in June 1968, and here one of the line's stalwarts, Ivatt Class 2 2-6-2T No 41241 leaves Haworth (home of the Brontë family) for Oxenhope on 3 April 1976. *(G T Heavyside)*

The doyen of the rescued Woodham locomotives, Fowler 4F 0-6-0 No 43924 and former Barry compatriot, Stanier 8F 2-8-0 No 8431 coast into Ingrow with a KWVR Enthusiasts' Day train on 3 April 1976.
(G T Heavyside)

GNR Atlantic No 990 *Henry Oakley*, on loan from the National Railway Museum, York, pilots oil-fired 0-6-0ST No 118 *Brussells*, a Hudswell Clarke built Austerity No 1782 in 1945, towards Ingrow station on the outskirts of Keighley on 11 September 1977.
(G T Heavyside)

The KWVR received widespread publicity from the feature film *The Railway Children*, and during a break in filming at Oakworth in May 1970 the three children, played by Jenny Agutter, Gary Warren and Sally Thomsett, stand alongside Manchester Ship Canal 0-6-0T No 67 (Hudswell Clarke No 1369 built 1919) which powered the 'London' train. *(J Gordon Blears)*

Above: Two unlikely companions! Barton Wright L&YR 0-6-0 No 52044 (the Green Dragon in *The Railway Children*) and USA 0-6-0T No 72 forge round Mytholmes curve with a Keighley—Oxenhope train on 26 March 1977. *(G T Heavyside)*

Top right: The valley reverberates as BR 9F 2-10-0 No 92220 *Evening Star* (which spent two years on Worth Valley metals) and WD Austerity 2-8-0 No 1931 approach Oxenhope with a Santa Claus special on 15 December 1974. No 1931 built by Vulcan Foundry, Newton-le-Willows, in 1945, operated initially in Holland and Sweden, before being cocooned by the Swedish State Railways in 1959 as part of its reserve stock. On returning home in 1973 it became the sole example in Britain, although BR once possessed 733 of the class. *(G T Heavyside)*

Right: The old gentleman's train engine in *The Railway Children*, London Transport 0-6-0PT No L89 (previously GWR No 5775) departs from Damems loop for Oxenhope on 21 July 1974. The loop was opened in 1971 to provide much needed extra line capacity at peak periods, and is controlled by the signalbox on the left, brought from Frizinghall on the MR Shipley—Bradford line. *(G T Heavyside)*

Left: Visitors to the Midland Railway Trust site at Butterley, north of Derby, admire LMS 'Jinty' 0-6-0T No 16440 coupled to an LMS dynamometer car on 29 August 1977. *(G T Heavyside)*

Right: Norwegian State Railways 2-6-0 No 377 *King Haakon VII* darkens the sky as it leaves Loughborough Central with a Main Line Steam Trust train for Quorn & Woodhouse on 6 July 1975. Services have subsequently been extended to Rothley. *(G T Heavyside)*

Below: More symbolic of motive power once seen on this former Great Central main line to Marylebone is Stanier Class 5 4-6-0 No 5231, here simmering gently at Loughborough on 10 October 1976. In May 1976 the engine was named *3rd (Volunteer) Battalion The Worcestershire and Sherwood Foresters Regiment*, one of the longest ever bestowed on a locomotive. *(G T Heavyside)*

Bottom right: Nameplate and regimental badge of No 5231. In BR days only four of the 842 members of this class carried names, but of 13 preserved six have been so honoured. *(G T Heavyside)*

3ʳᵈ (VOLUNTEER) BATTALION
THE WORCESTERSHIRE AND
SHERWOOD FORESTERS REGIMENT

Above: Standing at Shackerstone station, Leicestershire, are another batch of ex-industrial engines, where 1925-built Hunslet No 1493 0-4-0ST NCB No 11, a cut-down locomotive from Pye Hill Colliery, Jacksdale, Notts, passes Robert Stephenson & Hawthorns 0-6-0T No 7684 built 1951, from CEGB Nechells Power Station, Birmingham, in May 1973. Future plans include passenger services on the former LNW and MR Joint line to Market Bosworth, and eventually to Shenton.
(G D King)

Top right: Engines of Danish, French, German and Swedish origin can be seen alongside their British counterparts on the Nene Valley Railway, and on 16 July 1977 German built 2-6-2T No 064 305-6 is framed beneath Bulleid Battle of Britain Pacific No 34081 *92 Squadron* at Wansford. On the right is oil burning Swedish State Railways 2-6-4T No 1928.
(G T Heavyside)

Right: On the same day Swedish State Railways 2-6-2T No 1178 hauls Southern electric 4COR set No 3142 through a heavy storm at Ferry Meadows, forming the 15.10 Wansford—Orton Mere. *(G T Heavyside)*

Right: Horticulturist Alan Bloom has developed one of the finest steam museums within his own grounds at Bressingham Hall, Diss, Norfolk, where on 30 May 1974 Stanier Coronation class Pacific No 6233 *Duchess of Sutherland* dwarfs Orenstein & Koppel 0-4-0WT *Eigiau* on the 60cm gauge track. *Duchess of Sutherland* spent six years at Butlins holiday camp, Ayr, before arrival at Bressingham in 1971. *(G D King)*

Bottom right: In addition to providing pleasure for visitors, engines often find employment within the depots, as at the Chappel & Wakes Colne headquarters of the Stour Valley Railway Preservation Society where Robert Stephenson & Hawthorns 0-6-0T No 7597, built 1949, handles a rail mounted crane on 9 April 1977. *(G D King)*

Below: Hoping to work on the North Norfolk Railway in years to come is inside-cylinder 4-6-0 Class B12 No 61572, here seen in company with 0-6-0ST No 45 *Colwyn* (Kitson No 5470 of 1933) at Sheringham on 16 October 1977. *(G T Heavyside)*

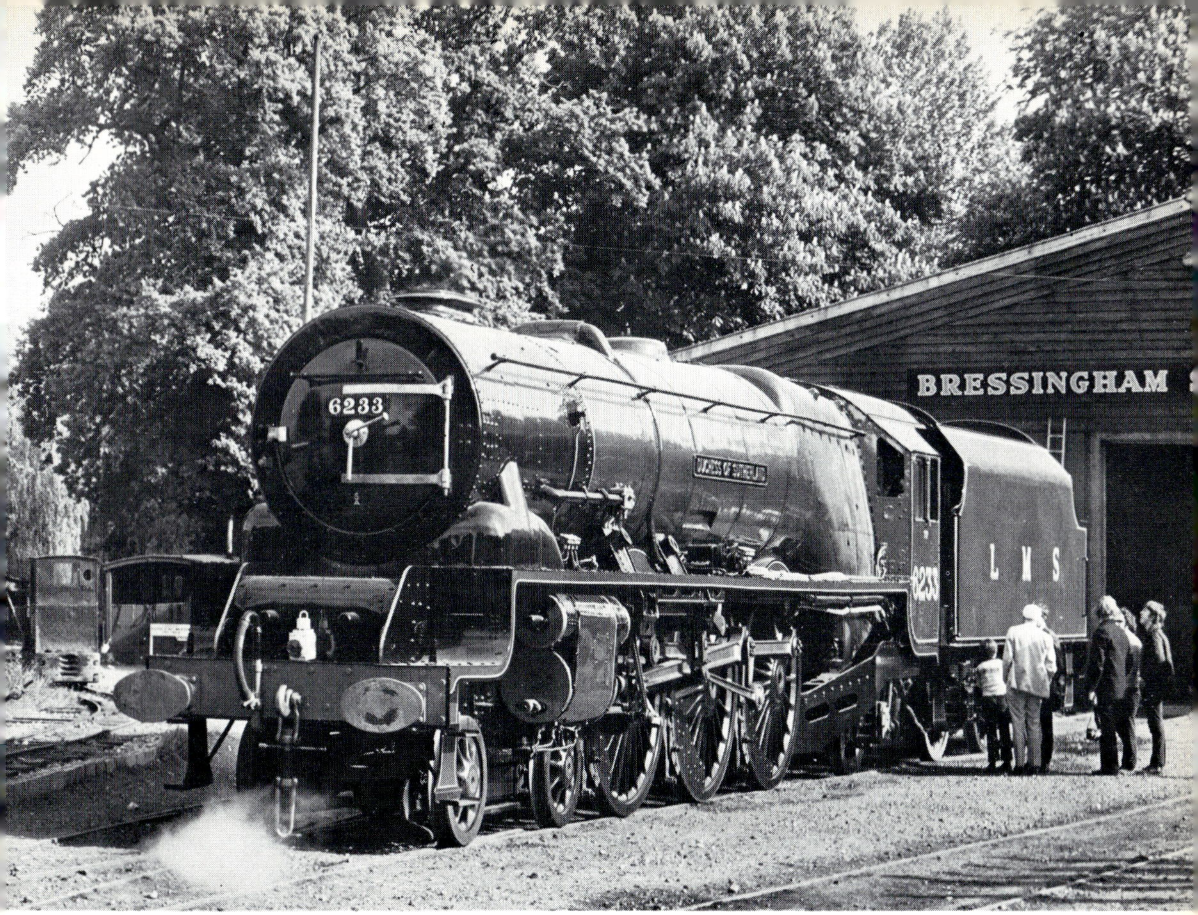
6233
BRESSINGHAM
L M S
6233

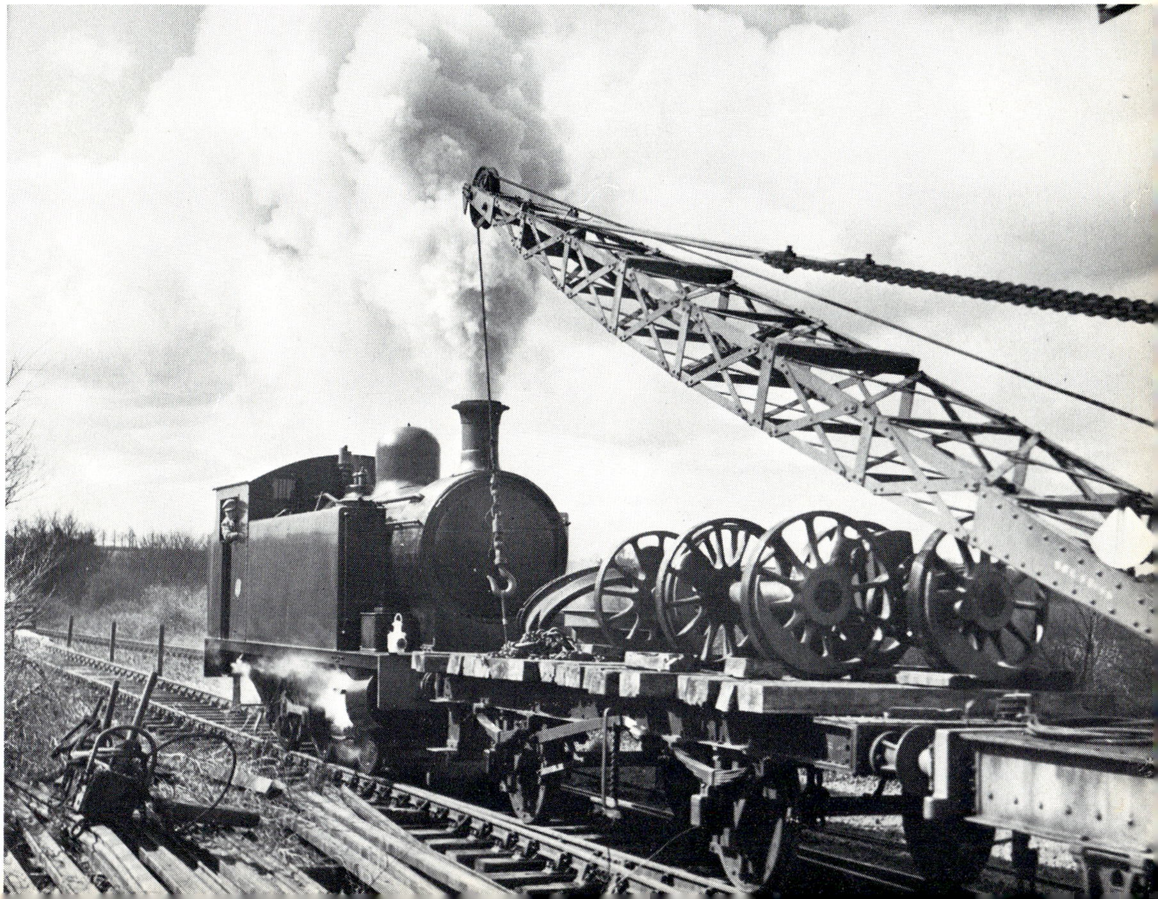

Above: Two years short of its centenary, rebuilt Stroudley 'Terrier' Class A1X 0-6-0T No 10 *Sutton* leaves Tenterden on the Kent & East Sussex Railway with the 15.30 to Rolvenden on 24 February 1974.
(D A Idle)

Right: Trains have been steaming north from Sheffield Park under the auspices of the Bluebell Railway for almost 20 years, and there are now long term plans to extend services from Horsted Keynes to East Grinstead. On 15 September 1974 BR Standard Class 4 4-6-0 No 75027 waits to leave Sheffield Park with the first train of the day. *(G T Heavyside)*

Top left: Two 1977 Bluebell acquisitions. SR Class U Mogul No 1618, recently arrived from the Kent & East Sussex Railway, shunts Bulleid austerity design Class Q1 0-6-0 No 33001, on loan from the National collection to the Bulleid Society, before the official handing over ceremony at Sheffield Park on 9 July 1977.
(G T Heavyside)

Right: LSWR Class 0415 4-4-2T No 488 built in 1885, at Horsted Keynes on 15 September 1974. No 488 along with two sisters monopolised the Lyme Regis branch for many years, long after the rest of the class had been scrapped. *(G T Heavyside)*

Left: Bulleid West Country Pacific No 21C123 *Blackmore Vale* makes light work of its task near Freshfield Halt on 9 July 1977, commemorating 10 years since steam was withdrawn from the Southern Region.
(G T Heavyside)

Bottom right: Wainwright Class P 0-6-0T No 323 *Bluebell* ambles through the Sussex Weald on 12 March 1972 with a train from Horsted Keynes. *(D A Idle)*

Below: One of 14 0-6-0Ts purchased by the SR from the US Army Transportation Corps in 1946 for use in Southampton Docks, No 30064, climbs to Freshfield from Sheffield Park on 16 May 1976. *(D A Idle)*

Outside the Great Western Society depot at Didcot (former BR shed 81E) 5600 class 0-6-2T No 6697 is flanked by 7200 class 2-8-2T No 7202 and GWR diesel railcar No 4 on 14 September 1974. *(G T Heavyside)*

Isle of Wight Steam Railway LSWR Class O2 0-4-4T No 24 *Calbourne* (a class which was the mainstay of Island motive power for many years) shunts three LBSCR coaches at Havenstreet on 14 July 1977.
(J Mackett)

The Winchester & Alton Railway commenced services on 30 April 1977 from Alresford to Ropley, but aims eventually to relay track to Alton. Here Maunsell SECR Class N 2-6-0 No 31874 *Aznar Line* enters Alresford on 10 July 1977. *(G T Heavyside)*

On the same day, carrying the shed plate of a former home, Exmouth Junction, *Aznar Line* eases past the Alresford distant signal. Rescued from Barry in 1974 the engine was named after the shipping company which sponsored the move to Alresford. *(G T Heavyside)*

Day-to-day running on a tourist steam railway

G M Kichenside, Commercial Manager, Dart Valley Railway

Within the last 30 years, ever since that handful of railway enthusiasts walked down a Welsh valley looking at what was left of the almost defunct Talyllyn Railway in the late 1940s wondering whether it was really possible for a group of amateurs to take over and restore a railway and run it with volunteers, railway preservation has grown by leaps and bounds. Such is the importance of some of these lines, particularly those in holiday areas or near large conurbations, that they now play a major role as tourist attractions. Indeed, no longer are they merely preserved railways, but tourist steam railways in their own right, running in the conditions of today and, moreover, have become big business. But with the number of passenger journeys on some lines heading towards 500,000 a year and cash turnover in excess of £¼million on certain individual lines, clearly there must be very careful and expert financial control over the whole operation of a tourist railway.

Right from the start the Dart Valley Railway was organised as a normal commercial company, originally with finance put up by business interests, to restore steam passenger services on part of the one-time Totnes-Ashburton branch. Hardly had that venture got under way when British Railways announced closure proposals for the Paignton-Kingswear section of the former GW Torbay main line, and this section was also taken over by the DVR. Thus the DVR has two isolated sections of line under its control, both in the highly scenic South Devon holiday areas. Indeed, it is the fact that these railways operate as attractions during the holiday season, when passenger potential is at its highest, which makes them successful, compared with operation in BR days as an all-year-round common carrier where summertime profits were turned into wintertime losses.

Administration is one area on private railways where overheads can be kept to a minimum with operation in the hands of a small management team covering all departments as on the Dart Valley Railway, and where all the full time staff have to be ambidextrous. Operating season engine crews, for example, will be found during the winter as fitters undertaking major overhauls on locomotives, while the permanent-way staff, apart from their normal duties in looking after the track, also take a hand in publicity distribution.

The tourist railway year really begins as one season comes to an end when the financial results, insofar as they are known, are reviewed, and the timetable for the following year is discussed to see whether any changes are needed. Timetable leaflets and posters are then prepared and copy sent to the printers, and advertising programmes are drawn up.

Major sidelines of tourist railways are the shops and cafes which are provided at the principal stations. Indeed on some railways the turnover from shop sales almost matches the revenue from ticket sales. The level of catering requirements, too, must be planned carefully since few people will want set meals but they will want something more than sandwiches.

It is of course the railway itself which is of prime importance, and timetables must take into account the passenger potential, the locomotives and stock available, and the train crews. On the Dart Valley it has always been the policy to keep the number of locomotives in steam to a minimum, in order to hold costs within reasonable bounds, to conserve the motive power, and to ensure good loadings on trains. Day-to-day servicing must also be considered, for coaches need cleaning, toilet equipment replenished, and windows and door handles polished. One feature of preserved railways is to show its rolling stock in its best possible condition, since a clean train is not only something of which the railway can be proud but it will be noticed by passengers. Even more will it be noticed—adversely—if it is dirty!

If I have looked rather more on the economic aspects of running tourist steam railways it is because they play such a vital part in their running. Steam locomotives are expensive machines to restore, operate and maintain, and if we are to keep steam running, not merely through the 1980s but towards the end of the century, the economic facts of life must be kept well to the fore.

The Paignton to Kingswear, Torbay & Dartmouth Railway, is unique in that it passed to independent ownership from BR without any break in services, the present owners, the Dart Valley Railway, taking control from 1 January 1973. In BR days the branch was classified 'double red' allowing the heaviest GWR engines — the Kings — to work through to Kingswear.

Right: The 18.15 Kingswear—Paignton clears the tree-lined exit from the 495yd Greenway Tunnel in the hands of GWR 2-6-2T No 4588 on 16 June 1975.
(T G Flinders)

Next page: Glorious Devon from the Dart Valley Railway. Collett Manor 4-6-0 No 7827 *Lydham Manor* hugs the River Dart at the approach to Britannia Crossing Halt travelling back to Paignton on 25 June 1975. *(G T Heavyside)*

Below: On 14 June 1977 No 4588, suitably adorned to celebrate the Silver Jubilee of Queen Elizabeth II, makes a vociferous start from Kingswear. *(G T Heavyside)*

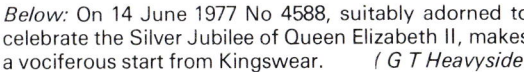

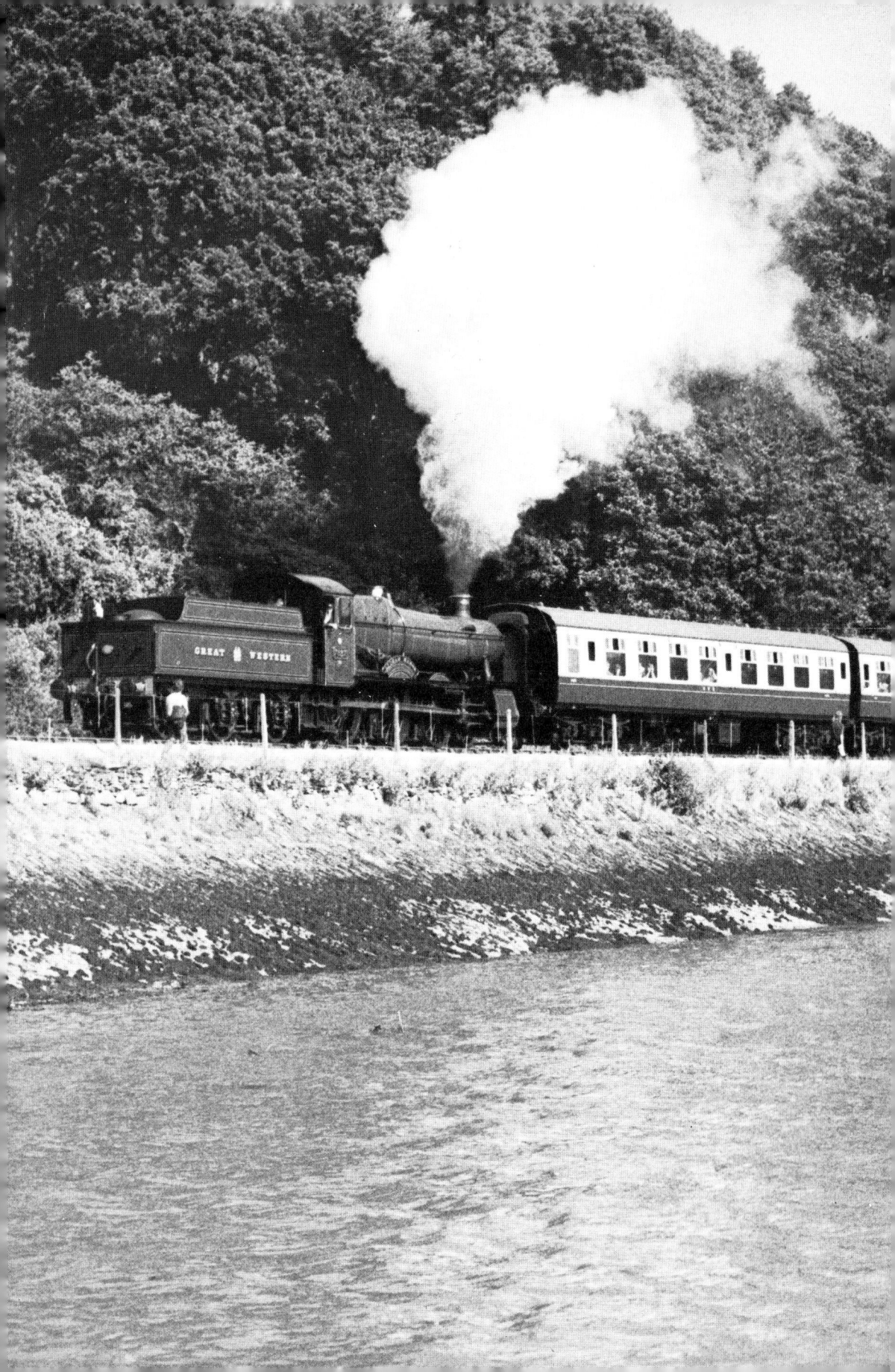

The Dart Valley Railway was officially reopened on 21 May 1969 by Lord Beeching, the former BR chairman during whose tenure of office the line was pruned from the BR network. A new platform has been built at Totnes, although without access from the town or BR station, since it is not possible for DVR trains to work into the BR station. At its northern end the branch is truncated at Buckfastleigh, new roadworks preventing access to Ashburton as originally envisaged.

Above: No 1420, now named *Bulliver*, takes water at Buckfastleigh on 14 June 1977. The leading vehicle is one of the two former Devon Belle observation cars, from which passengers obtain a panoramic view of the lush scenery. Its sister car went to the USA with the Flying Scotsman tour in 1969/70. *(G T Heavyside)*

Below: In October 1969 GWR 0-6-0PT No 6412 halts at Nappers Crossing while the fireman opens the level crossing gates. The engine has since been transferred to the West Somerset Railway. *(G T Heavyside)*

Collett 0-4-2T No 1420 makes sedate progress through the rich rolling countryside near Staverton Bridge, with a Totnes—Buckfastleigh train on 22 July 1973. The River Dart can be glimpsed through the trees in the left foreground. *(R E B Siviter)*

When the Dart Valley Railway first opened the line from Buckfastleigh trains were worked push-pull fashion. Here No 1420 pauses at rose-bordered Staverton Bridge in 1970. *(T G Flinders)*

Top left: This typical GWR scene at Minehead appears almost timeless as 0-6-0PT No 6412 awaits departure for Blue Anchor on 21 June 1977. The West Somerset Railway insignia and the Flockton Flyer headboard — the engine being the star attraction in this television serial — bring us up to date. *(G T Heavyside)*

Above: The East Somerset Railway owes much to the famous artist David Shepherd, as does the steam preservation movement, and at Cranmore on 24 April 1977 his BR 9F 2-10-0 No 92203 *Black Prince* gazes at 'Jinty' 0-6-0T No 47493 steaming by with a couple of brake-vans. Passenger trains are planned from here along the former GWR Cheddar Valley line for about 1½ miles towards Shepton Mallet. *(P J Fowler)*

Left: The tremendous amount of work required to restore some engines to working order is illustrated by this photograph of GWR 2-6-2T No 5542 at Bishops Lydeard on the same day. WSR trains have run to here from Minehead, the ultimate goal being Taunton.
(G T Heavyside)

Above: A far cry from charging the spoil tip at Hafodyrynys Colliery, Pontypool, Hunslet Austerity 0-6-0ST No 3810 *Glendower*, built in 1954, pulls a train carrying Santa Claus into Bitton station, Bristol Suburban Railway, on 11 December 1977. This engine, and two other austerity 0-6-0STs, moved to the Dart Valley Buckfastleigh line in March 1978. *(P J Fowler)*

Right: At the Dean Forest Railway Society, Parkend depot, GWR Prairie tank No 5541 undergoes a steam test on 28 March 1976, while another ex NCB Hunslet Austerity No 3806 of 1953 passively looks on. The society has a second site at nearby Norchard.
(G T Heavyside)

On 16 April 1977 'Jinty' 0-6-0T No 47383 stands outside the recently erected shed in Bridgnorth yard, Severn Valley Railway, which now provides much needed covered accommodation. On the left is Manning Wardle 0-6-0ST No 2047 *Warwickshire*, built 1926.
(G T Heavyside)

The next day, near the other extremity of SVR operation, Longmoor Military Railway 2-10-0 No 600 *Gordon* battles up to Bewdley Tunnel with a train from Bridgnorth. *(G T Heavyside)*

Planning for the future

D C Williams, Director, Severn Valley Railway Co Ltd

All preserved railways must of necessity plan ahead, and though this article attempts to set out what is involved in future planning for the Severn Valley Railway, the problems encountered are common to most, if not all, of the other preserved railways. Though running regular train services over 12½ miles of track, there are only 17 full-time staff on the Severn Valley, the rest of the work being performed by between 100 and 200 unpaid volunteers, working mostly at weekends. Since the great burden of day-to-day working devolves upon the paid staff, it follows that future planning is generally the responsibility of unpaid people, mostly at the monthly meetings of the Directors of the Holdings and Guarantee Companies, the former responsible for finance and overall policy, the latter for operating the railway. Approximately 15 people sit around the table. A smaller 'Planning Department' meets from time to time, and has produced some very useful reports for the Directors.

Despite the ever-present commitment to run a very long railway, most people connected with the SVR want to extend services to Kidderminster to give a permanent connection with British Rail. Unfortunately we are being asked to purchase a suitable terminus—Kidderminster Goods Shed—from the National Freight Corporation well before we know whether the 1,500yd section of track from Foley Park to Kidderminster will ever be ours. At present it is used by BR for traffic to the British Sugar Corporation factory at Foley Park. Like so many things in railway planning, in 2 or 3 years time the modern historians will all be able to tell us whether we made the right decision in this 'cart before the horse' situation. The Goods Shed will cost upwards of £30,000 if we decide to buy it.

We now pass on from possible expansion to defence of our present railway. Services only commenced from Bridgnorth to Hampton Loade in 1970 subject to the SVR agreeing to pay for a rail-over-road bridge at the point where a proposed southern by-pass for Bridgnorth would cross the line. A by-pass bridge fund is in existence but the SVR Board keep a watching brief on this expensive threat to the line's future. Estimated cost of the bridge is £300,000, but the cost of the whole by-pass project inflates annually and pushes the scheme further into the future.

In these difficult times, financial matters assume even greater importance. It is necessary to plan future improvements in the service we give to the public, and this capital for new projects has to be raised by means of share issues. Authorised share capital of the SVR Holdings Company is presently £550,000, though only £350,000 has been taken up by approx 1,500 shareholders. The timing of a future issue to raise up to £200,000 is most important, and must be coupled with the plans for Kidderminster Goods Shed, Bridgnorth by-pass bridge, and other important capital items. Most members consider that the provision of improved catering facilities for Bridgnorth, and the erection of buildings for the repair and storage of carriages at Bewdley are the two most important requirements for the future.

Despite a huge effort to publicise the railway, it must be admitted that there is room for improvement in the SVR traffic figures. Plans are made by a small committee before the start of each season, based on the results of the previous year's running, the object being to provide enough trains to satisfy the vast majority of our visitors, but not so many trains as to be running around very lightly loaded. This balance is a very fine one. The annual figures of approx 100,000 passenger tickets sold results in a near break-even situation at present; 120,000 passengers would give the railway a healthy profit, and 150,000 would put a permanent smile on the faces of all SVR supporters. The publicity budget thus assumes great importance. How many TV commercials can we afford (at over £500 each) and how effective are they compared with other forms of publicity?

The foregoing has touched only briefly on some of our planning problems and a whole book could be devoted to this side of operations. Much depends on attracting working members as well as the passengers, and with no lack of opportunities for skills of all kinds, assistance is always welcome!

Shortly after leaving Bridgnorth Stanier 8F 2-8-0 No 8233 rumbles over Oldbury viaduct on 16 April 1977.
(G T Heavyside)

On the same day on its 14½ mile journey from Foley Park to Bridgnorth GWR 2-6-2T No 4566 crosses Hay Bridge, approaching Eardington. *(G T Heavyside)*

Ivatt Mogul No 46443, still disguised as Austrian State Railways No 60 116 (see page 96) for its part in the film *The Seven Per Cent Solution* heads north from Bewdley on 16 May 1976. In the foreground is the track-bed of the closed Wyre Forest line which diverged from the SVR at this point. *(G T Heavyside)*

Hunslet Austerity 0-6-0ST No 193 (works No 3793, built 1953) passes the picturesque SVR Arley station with an Enthusiasts' Day demonstration goods train on 12 September 1976. *(G T Heavyside)*

Left: Standard Class 4 2-6-4T No 80079 blasts out of Bewdley Tunnel with an SVR Spring Steam Gala train for Bridgnorth on 17 April 1977. With no run round facilities at Foley Park trains require an engine at each end when traversing this section from Bewdley, although it is normally only used on special occasions.
(G T Heavyside)

Above: During one of its rare public appearances LNER Class A2 Pacific No 532 *Blue Peter* eases past Tyseley station while working a short shuttle service from Birmingham Railway Museum on 6 October 1974.
(R E B Siviter)

Above: The fisherman's concentration is momentarily disturbed by Hawthorn Leslie 0-4-0ST No 2780 *Asbestos*, built 1909, hauling former BR Gloucester dmu trailer car No E56301 along the Chacewater Light Railway, Brownhills, on 18 August 1974.
(G T Heavyside)

Right: Another 0-4-0ST, this one built by Andrew Barclay in 1927 as No 1927, is put through its paces on a short section of track outside Bury Transport Museum on 31 July 1977. *(G T Heavyside)*

Below: Foxfield Light Railway Bagnall 0-4-0ST No 2623 of 1940 *Hawarden* summons all its energy to lift one coach up the ferocious grade from Foxfield station, Dilhorne, on this one-time colliery line to Blythe Bridge, near Stoke-on-Trent, on 2 May 1976. *(G T Heavyside)*

Deutsche Dampflokomotiven bei Carnforth. During a continental week-end on 20 March 1976 East German 0-6-0T No 80 014 poses alongside West German Pacific No 012 104-6 outside Steamtown. A French Pacific is also housed at this former BR shed, together with many ex-BR and industrial locomotives. *(G T Heavyside)*

Above: Ex MOD Shoeburyness Hunslet Austerity 0-6-0ST (No 3794 of 1953) *Cumbria* awaits departure time at Lakeside on the Lakeside & Haverthwaite Railway on 4 July 1976. Many trains on this former Furness Railway branch, now isolated from BR, connect here with Sealink vessels on Lake Windermere.
(G T Heavyside)

Top left: LMS Fairburn 2-6-4T No 2085, masquerading in Caledonian Railway colours, makes a spirited start from Haverthwaite on 21 September 1974.
(G T Heavyside)

Left: Sister engine No 2073 clatters past Chapel Crossing on 22 July 1973 with the 15.32 Haverthwaite — Lakeside. *(G T Heavyside)*

Return to the main line

To the chagrin of many people, British Rail announced that from the end of 1967 no privately-owned steam locomotives would be allowed to haul trains over its tracks, except LNER A3 Pacific No 4472 *Flying Scotsman* whose then owner, Alan Pegler, held a contract which enabled the engine to work specials until December 1971. With *Flying Scotsman* departing for America in September 1969, 13 months after the end of steam on ordinary workings, a totalitarian state of diesel and electric power was thus created on BR. During these austere times, if it was necessary for steam to travel over the main line they normally suffered the indignity of being towed by a diesel, even when travelling to open days at BR depots, whose organisers realised their potential in attracting clientele.

It seemed ironic that the iron horse should be tethered to light railways and private sidings, sometimes with barely room to exercise the pistons, and there was a widespread feeling that steam should again be allowed to show its true colours on the main line. There was no lack of precedent for preserved engines heading specials on BR, the trend being set in 1953 (following a foray over LNER tracks by Stirling Single No 1 in 1938) when GNR Atlantics No 990 *Henry Oakley* and No 251 were taken out of York railway museum and hauled a train from Kings Cross to Doncaster, returning to London the following week, starting from Leeds. Over the years BR restored other engines to their pre-grouping liveries for special workings, including MR compound 4-4-0 No 1000 and LSWR Class T9 4-4-0 No 120, while the historic GWR 4-4-0 No 3717 *City of Truro* (running as No 3440) was also resurrected from York Museum. In the dying years of the steam era some privately-owned locomotives followed suit.

With a view to reviving these golden days the Association of Railway Preservation Societies formed a 'Return to Steam Committee' in 1969, hoping to realise its aims by the 150th anniversary of the Stockton & Darlington Railway in 1975. At times the bastions of the BR Board appeared impregnable and many arguments were advanced why steam could not be allowed, although often they had little foundation and were effectively countered in the railway press. The barriers were, however, breached much sooner than expected when the dynamic Peter Prior, managing director of H P Bulmer Ltd, whose company had custody of GWR King Class 4-6-0 No 6000 *King George V* at its Hereford factory, arranged a series of proving runs with the engine to assess the feasibility of future steam specials. It was a momentous day on 2 October 1971 as *King George V* steamed south towards Newport with the Bulmers cider train of ex-Pullman cars and two BR coaches, and then on to Tyseley via Didcot. During the following week the train travelled to Kensington before returning to Hereford via Swindon, with thousands lining the route to witness the giant pass, and once more relish the sight of main line steam.

The tour can only be described as a resounding success and in 1972 BR decided to allow a limited number of steam-hauled excursions over five widely dispersed secondary routes totalling 301 miles, each having locomotives stabled nearby and turning facilities available at each end in the form of triangles or turntables. Twenty-three engines were listed as suitable for hire to BR at a nominal sum for this purpose although, not unnaturally, very stringent conditions were laid down regarding mechanical condition, insurance cover, etc, while the preservation bodies had to be responsible for all servicing. The position was to be reviewed at the end of the season's running.

In 1972 eleven trains had steam at the head and with no real problems encountered, and the added bonus of a small profit for BR, the pattern was set for future years. In fact the following year an additional 460 miles were approved for steam running and at the time of going to press the total is about 900 miles, although some lines after a short spell of activity have been deleted from the list. In 1975 as a one off venture in connection with the Rail 150 celebrations for the Stockton & Darlington 150th anniversary in the North East, BR allowed four trains with steam power on the East Coast main line between York and Newcastle, three starting with steam at Sheffield, and also others on the Battersby to Whitby line. An earlier unique event was the Maidenhead—Marlow branch centenary celebrations on 15 July 1973, when two engines from the Great Western Society depot at Didcot shared services for the

Above: In the late 1950s the Scottish Region restored four veteran locomotives to pre-grouping condition for main line excursions, before transfer to Glasgow Transport Museum in 1966. Here CR 4-2-2 No 123 and GNSR 4-4-0 No 49 *Gordon Highlander* leave Newton Stewart with a railway societies' special on 23 June 1962. *(Derek Cross)*

Below: On 2 October 1971 GWR King class 4-6-0 No 6000 *King George V* regally hauls the Bulmers cider train at Pilning on the climb from the Severn Tunnel with the maiden 'Return to Steam' trip from Hereford to Tyseley. The bank of photographers on the left represent but a minute proportion of those who thronged the lineside. The bell carried over the front buffer commemorates another pioneering venture when the locomotive attended the Baltimore & Ohio Railroad centenary celebrations in September/October 1927. *(P J Fowler)*

day with 0-4-2T No 1450, normally based on the Dart Valley Railway.

Since 1974 steam specials have averaged 26 a year, although regrettably some scheduled trips have been cancelled, while 30 locomotives have tasted main line service again, the most notable being LNWR 2-4-0 No 790 *Hardwicke* which performed heroically in the 1895 Railway Race to Aberdeen on the Crewe to Carlisle section. With regional boundaries counting for nothing in this new epoch the scene has often resembled the 1948 locomotive exchanges (as was the case

Ready for main line duty again. Thompson Class B1 4-6-0 No 1306 *Mayflower* shortly after passing Lostock Junction, near Bolton, on 19 March 1975 returning to Steamtown, Carnforth, via Hellifield, following attention at BREL Horwich Works. The name *Mayflower* was originally conferred on sister engine No 61379, 1306 taking the name in 1971. *(G T Heavyside)*

for a short time in the mid 1960s) with LNER engines travelling to Didcot, Newport, and Sellafield, and SR Merchant Navy class Pacific No 35028 *Clan Line* visiting Chester and Stratford, amongst other previously undreamed of journeys. However from 1976 the cavorting round the country came to an end when BR, in approving steam running until at least 1979, decreed that engines in future would normally be confined to the routes adjacent to their depots. This move was to facilitate administration and enable inspectors, etc, to familiarise themselves with certain engines. The locomotive owners and operating societies have also benefited for costly light engine movements have been eliminated, although some would like to see this rule relaxed.

In addition to trains organised by the recognised railway societies steam has been used on other charter specials, and also to draw exhibition trains whose sponsors have realised the magnetic power of steam on the general public.

Another notable milestone, 19 October 1974, saw Great Western Society 4-6-0s No 7808 *Cookham Manor* and No 6998 *Burton Agnes Hall* proudly hauling a complete set of ex-GWR coaches, beautifully restored in chocolate and cream, from their home shed at Didcot to Stratford and Tyseley. This was the first time that a whole train of privately preserved stock worked over BR, and was the forerunner of other trips from Didcot, and also some with a similarly restored GWR collection normally found at work on the Severn Valley Railway. On many occasions private saloons, some of ancient vintage and renovated to pristine condition, have been used, contrasting sharply with standard BR coaches. These vehicles are subject to rigorous examination and testing before being allowed on the main line.

Over the past few years it has been a stirring sight for many to witness a steam locomotive at the head of a heavy train, especially while clambering such heights as Giggleswick, Hatton, and Llanvihangel when maximum exertion is required. The charisma of such scenes is often indelibly printed in the mind of the beholder but for how long the spectacle will be re-enacted is impossible to forecast, and depends on various factors, not least being that of finance and the continuing goodwill of BR. Hopefully it will be for many years, giving young people, who providence decided were born too late to sample steam on normal BR services, the occasional opportunity to see the iron horse with the reins relaxed, and for those of more mature years welcome refreshment for the memory.

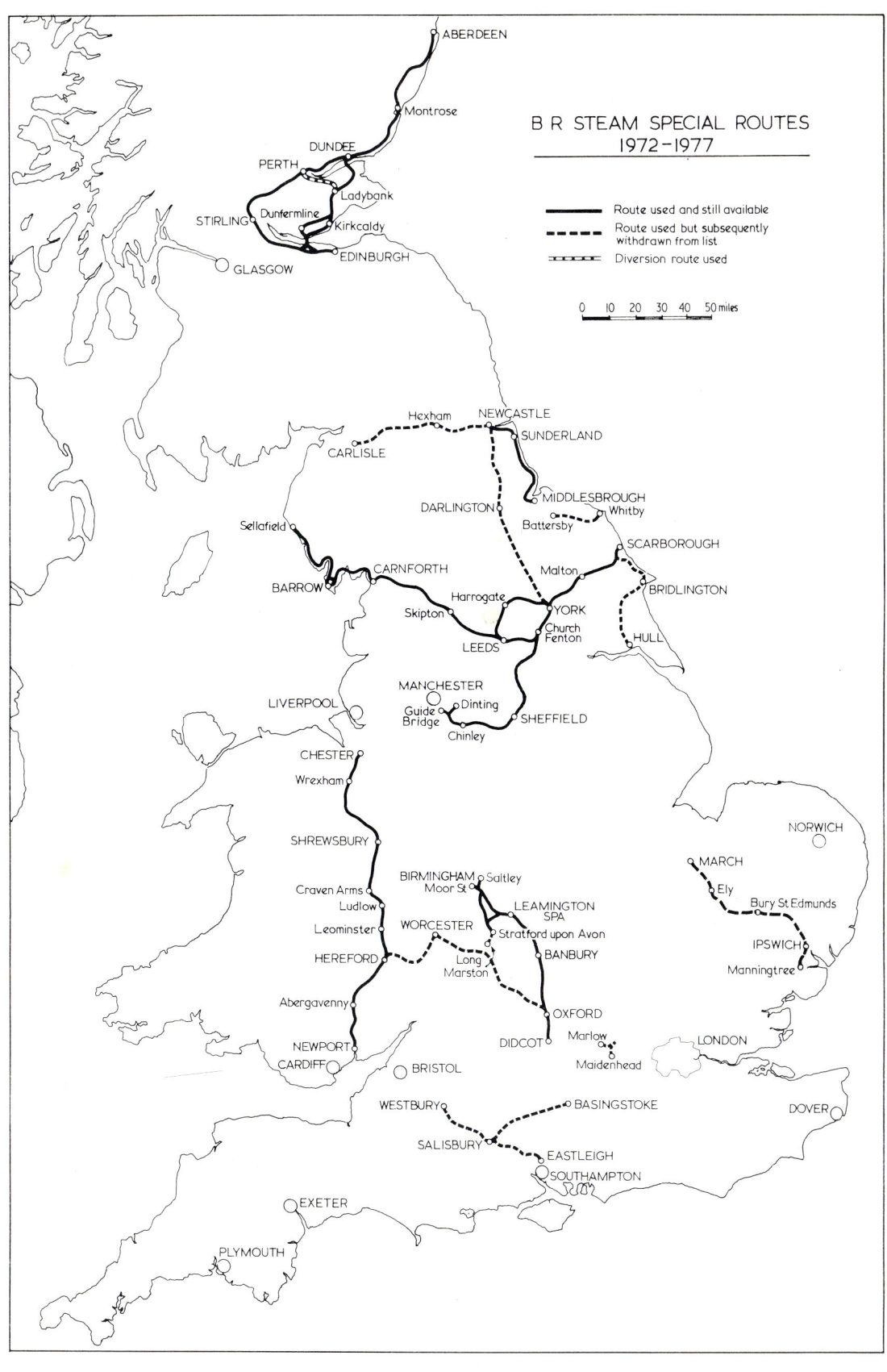

Left: Named after the designer, LNER A4 Pacific No 4498 *Sir Nigel Gresley* thunders out of Aberdeen past Ferryhill on 22 June 1974. *(J S Whiteley)*

Below: Sister engine No 60009 *Union of South Africa*, in BR guise, heads south from Perth on 7 June 1975. *(J S Whiteley)*

Right: K1 class 2-6-0 No 2005 pulls away from Gateshead on 4 September 1977 with a Newcastle to Whitby via Sunderland train, which it will haul as far as Middlesbrough. *(G T Heavyside)*

Below: Sir Nigel Gresley stands at Carlisle station after arrival from Newcastle as BR Class 50 No 443 enters with an Anglo-Scottish express on 6 October 1973.
(G T Heavyside)

Above: Standard Class 9F 2-10-0 No 92220 *Evening Star* was the last steam locomotive built for BR in 1960; withdrawn from stock only five years later, the engine now normally resides in the National Railway Museum, York, but makes occasional main line appearances. Here *Evening Star* rounds the sinuous curves through the Howardian Hills at Kirkham Abbey en route to Scarborough on 30 October 1976. *(D Rodgers)*

Top right: Gresley A3 Pacific No 4472 *Flying Scotsman* makes its exit from Scarborough past Falsgrave signalbox on 12 June 1976. On the right a Brush Class 47 waits to follow the same path to York. In 1934 *Flying Scotsman* became Britain's first steam locomotive to reach an authenticated speed of 100mph.

(G T Heavyside)

Right: Gresley A4 Pacific No 19 *Bittern* whistles through Driffield between Scarborough and Hull on 21 April 1973. The route was used only for steam on this occasion although engines working into Scarborough travel to Filey Holiday Camp to turn on the triangle.

(P J Fitton)

Above: On 24 April 1976 LNWR Precedent class 2-4-0 No 790 *Hardwicke* and MR compound 4-4-0 No 1000 cross the River Nidd at Knaresborough with a train from York to Carnforth, where they were scheduled to stay at Steamtown for two months on loan from the National Railway Museum, for working specials in the North West. Unhappily the compound subsequently failed and was unable to cover its other rostered duties.
(G T Heavyside)

Top right: LMS Black 5 4-6-0 No 5305 enters Wharfedale through the castellated portal of Bramhope Tunnel with a Leeds to Scarborough via Harrogate special on 30 April 1977. Following cessation of BR steam No 5305 was consigned to Albert Draper & Son Ltd but fortunately was reprieved from the torch by the Hull scrap-merchants, and subsequently restored to its early splendour by local enthusiasts. *(D Rodgers)*

Right: LNER B1 class 4-6-0 No 1306 *Mayflower* approaches Marsh Lane cutting, Leeds, on the direct route to York via Church Fenton with a private charter from Carnforth on 12 June 1977. *(G T Heavyside)*

Organising a steam special

C B Phillips, Secretary, Essex Locomotive Society Ltd

Whenever the Essex Locomotive Society engine, SR Class S15 4-6-0 No 841 *Greene King* speeds across East Anglia with a rake of BR coaches in tow, few realise the tremendous amount of time and effort that goes into making it all possible. Those enjoying the locomotive's every beat and hiss at first hand, and the clicky-ti-click of steel rail underneath—where welded track has not yet replaced jointed rail—let alone the hundreds gathered at the lineside, spare little thought for the administrative load carried by a small band.

Planning a special starts months in advance, when the directors first formulate a proposal to put before BR, having first consulted the Steam Locomotive Operators Association who co-ordinate the trains to ensure that those projected for the same area do not clash. The only route available for *Greene King* was from Manningtree to March via Bury St Edmunds, and to attract potential participants from London our trains started from Liverpool Street. With the steam section early in the journey we must look for a suitable destination which will prove popular with our clientele, and ELS specials have visited Loughborough (Main Line Steam Trust), York (National Railway Museum) and Dinting, including a second steam-hauled section to Sheffield behind Jubilee class 4-6-0 No 5690 *Leander*.

Once the tour outline has been finalised with BR and their fees known for providing the coaching stock, a buffet car if required, the crews and locomotive inspector, the delicate task of deciding the fare must be faced, not forgetting the many sundry items which have to be financed, such as coal and watering charges, advertising, insurance, etc. The price must of necessity be high and sometimes appears exorbitant when compared to similar BR excursions, but even so is often barely adequate to cover the many contingencies. It is thus mainly railway enthusiasts who will support a steam trip, and with this in mind, suitable advertising space is taken in the railway journals.

As the big day draws nearer detailed planning continues; train stewards are appointed, servicing facilities for *Greene King* arranged, the engine inspected by BR and any necessary adjustments made, amongst a multitude of

Left: Nocturnal view of SR Class S15 4-6-0 No 841 *Greene King* at Ipswich following arrival from Norwich on 4 October 1977, in preparation for an Essex Locomotive Society special on 15 October. *(G D King)*

Right: Unfortunately even the best laid plans occasionally go awry, as on the misty morning of 15 October, due to steaming problems, *Greene King* required the assistance of BR diesel No 31 125, and was only able to work as far as Ipswich. It is seen here crossing the River Stour at Manningtree.
(G T Heavyside)

other tasks. Attention to detail is vital, mistakes on some fronts could spell total disaster.

During this period of intense activity the office of booking clerk becomes no sinecure, with a porch full of mail, a red hot telephone, and everyone demanding prompt attention. Perhaps at times the job becomes somewhat frustrating as he endeavours to complete his carefully drawn seat plan and simultaneously satisfy all customers. These include Mr Smith who must travel in the front coach (in both directions) in order to record the engine's vocal delights on tape, Mr Jones who desires to photograph the engine from the fourth coach, and Mr Brown who demands a facing seat on the left hand side as he wishes to assess progress and check speeds utilising stop watch and mile posts. Complications arise when a traveller from London wishes to sit next to a friend who will board at Ipswich, but whose application is not to hand anyway. The task is made no easier by the number of irregularly drawn cheques received, over and under payments, illegible writing, and even suggestions that the whole itinerary should be changed, all of whom must have polite replies.

Shortly before the day, BR will forward detailed timings so that these can be included in a brochure, together with a route description and other relevant information, to be handed to all participants. The specially printed tickets can now be posted along with a list of start and return times from each departure station. The engineering staff prepare to carry out the final servicing and then the lighting up. This means a complete night at Ipswich depot.

The great day dawns and while committee members may breathe a sigh of relief that shortly the fruits of much labour will be manifest, there is still a busy day ahead to ensure the smooth running of all aspects and keeping everyone happy. There is also the opportunity, not to be missed, of publicising the Society, and through book and souvenir sales raise a little extra revenue.

It is only later when all administrative tasks have been completed that there is time to reflect and evaluate the exercise. Lessons will no doubt have been learnt and mental notes made of points to remember in future. The financial director's balance sheet will be carefully scrutinised, hoping this will show a welcome profit to augment the funds required to keep the engine in working order. There comes too the satisfaction of knowing that through the Society's efforts a further contribution has been made to another chapter in the life of steam.

Left: BR Standard Class 9F 2-10-0 No 92203 *Black Prince* skirts the western edge of Salisbury Plain near Warminster en route to Eastleigh from Westbury on 20 April 1975. This class was an undoubted success story of latter day steam locomotive development, and while designed for heavy freight haulage BR soon found them equally at home on passenger trains, a fact taken advantage of in the preservation era. *(P J Fowler)*

Below: Later on the same journey *Black Prince* is seen on the climb to Salisbury Tunnel. *(T G Flinders)*

Bottom left: Back on its old haunts, the LSWR Waterloo—Exeter main line, rebuilt Bulleid Merchant Navy 4-6-2 No 35028 *Clan Line* heads towards Grateley, returning to Basingstoke from Westbury on 27 April 1974. These lines have subsequently been deleted from the list of approved routes. *(T G Flinders)*

On an overcast 26 October 1974, at Bodicote, in the Cherwell Valley, south of Banbury, rebuilt Merchant Navy Pacific No 35028 *Clan Line* heads for Stratford-upon-Avon. *(G T Heavyside)*

With the first complete train of GWR restored stock to grace BR metals 4-6-0s No 7808 *Cookham Manor* and No 6998 *Burton Agnes Hall* climb to Wilmcote from Stratford on 19 October 1974. *(J S Whiteley)*

Hall class 4-6-0 No 5900 *Hinderton Hall* piloted by Modified Hall class 4-6-0 No 6998 *Burton Agnes Hall* pull away from a signal check at Small Heath in the Birmingham suburbs, returning to their home depot at Didcot on 15 May 1976, with a Great Western Society special. *(G T Heavyside)*

Gresley V2 2-6-2 No 4771 *Green Arrow* storms up Hatton bank with a Didcot to Tyseley working on 1 July 1973. *(J S Whiteley)*

Left: Regrettably steam is again but a memory on the Oxford to Hereford line, as it allowed through running between two other routes. With the last trip to thread the Cotswolds *Cookham Manor* and *Burton Agnes Hall* cross the Severn at Worcester on 14 June 1975.

(R E B Siviter)

Bottom left: Sunday 29 May 1977 was a sad day for British enthusiasts as it provided the last opportunity to savour GWR Castle class 4-6-0 No 4079 *Pendennis Castle* on a main line tour before shipment to its new owners in Western Australia, Hamersley Iron Pty Ltd, whose 240 mile mineral line will in future echo Swindon practice on tourist trains. The engine originally found fame when it ran over LNER metals in the 1925 locomotive exchanges. Here *Pendennis Castle*, appropriately bearing the headboard 'The Great Western Envoy' passes the closed station at Aynho, wrong line due to engineering work, on the outward journey from Birmingham to Didcot, with the valediction special. The viaduct seen above the last coach carries the down line of the 1910 direct route to Paddington via Princes Risborough. *(G T Heavyside)*

Below: From the last batch of Castles, No 7029 *Clun Castle,* one of seven preserved examples to remain on home soil, threads the university city of Oxford and makes for Didcot on 4 October 1975. *(G T Heavyside)*

Right: GWR splendour relived. To commemorate 50 years of the King class 4-6-0s, No 6000 *King George V* hauled a set of Severn Valley Railway restored GWR carriages over the North and West route between Newport and Shrewsbury, here cautiously passing Pontypool on 3 July 1977. *(G T Heavyside)*

Bottom right: Far from its Dinting Railway Centre home, double chimney Jubilee 4-6-0 No 5596 *Bahamas* roars by the closed station platform at Bromfield, north of Ludlow, running from Hereford to Shrewsbury on 14 October 1972. *(P J Fitton)*

Below: Two overseas travellers in harness. On its first main line special since returning from America, the pride of the LNER, A3 Pacific No 4472 *Flying Scotsman*, leads the 1927 visitor to the USA, *King George V*, up the 1 in 95 to Llanvihangel summit with the north-bound Atlantic Venturers Express on 22 September 1973.
(P J Fitton)

Above: Against a backdrop of The Long Mynd, GWR Castle class 4-6-0 No 4079 *Pendennis Castle* accelerates towards Shrewsbury, from the slack through Church Stretton station on 6 April 1974. *(G T Heavyside)*

Above right: Returning from Chester with the train seen below *Princess Elizabeth* rides high over the graceful 19 arch viaduct across the Dee Valley at Pentre.
(G T Heavyside)

Below: King George V pounds away from Shrewsbury towards Chester, past Coton Hill sidings, with the SVR GWR train on 23 April 1977. *(G T Heavyside)*

Right: Stanier Princess Royal Pacific No 6201 *Princess Elizabeth* passes Baystonhill, on the long taxing climb from Shrewsbury to Church Stretton, on 24 April 1976.
(J S Whiteley)

Above: In the suburbs of Sheffield three-cylinder Jubilee class 4-6-0 No 5690 *Leander* approaches Totley Tunnel (3 miles 950 yards long, Britain's second longest) on the Hope Valley route through the Pennines, having just left the Midland main line at Dore on 16 June 1974. *(G T Heavyside)*

Top right: Climbing away from Chinley towards the Peak District on 30 October 1976, *Leander* belies the fact that it lay rotting at Barry for eight years before being rescued in 1972. *(G T Heavyside)*

Right: Slicing through the back-bone of England, *Leander*, assisted by LMS Black 5 4-6-0 No 5305, about to enter Cowburn Tunnel (2 miles 182 yards), returning to Guide Bridge from Sheffield on 24 September 1977. *(G T Heavyside)*

Gresley A3 4-6-2 No 4472 *Flying Scotsman* hauls the Pioneer hi-fi exhibition train through Newlay cutting, between Leeds and Shipley, at the start of the train's national tour on 25 July 1976. *(G T Heavyside)*

On 16 June 1974 V2 class 2-6-2 No 4771 *Green Arrow* hurries through the rain towards Leeds, past Kildwick level crossing, between Skipton and Keighley. *(G T Heavyside)*

Under the shadow of Ingleborough LNER Class B1 No 1306 *Mayflower* and LMS Class 5 No 45407 storm along the 'Little North Western' between Clapham and Giggleswick on 16 October 1976. *(G T Heavyside)*

David and Goliath in tandem. The diminutive LNWR 2-4-0 No 790 *Hardwicke* and BR 2-10-0 No 92220 *Evening Star* climb through Clapham cutting, returning home to York, on 19 June 1976. *(G T Heavyside)*

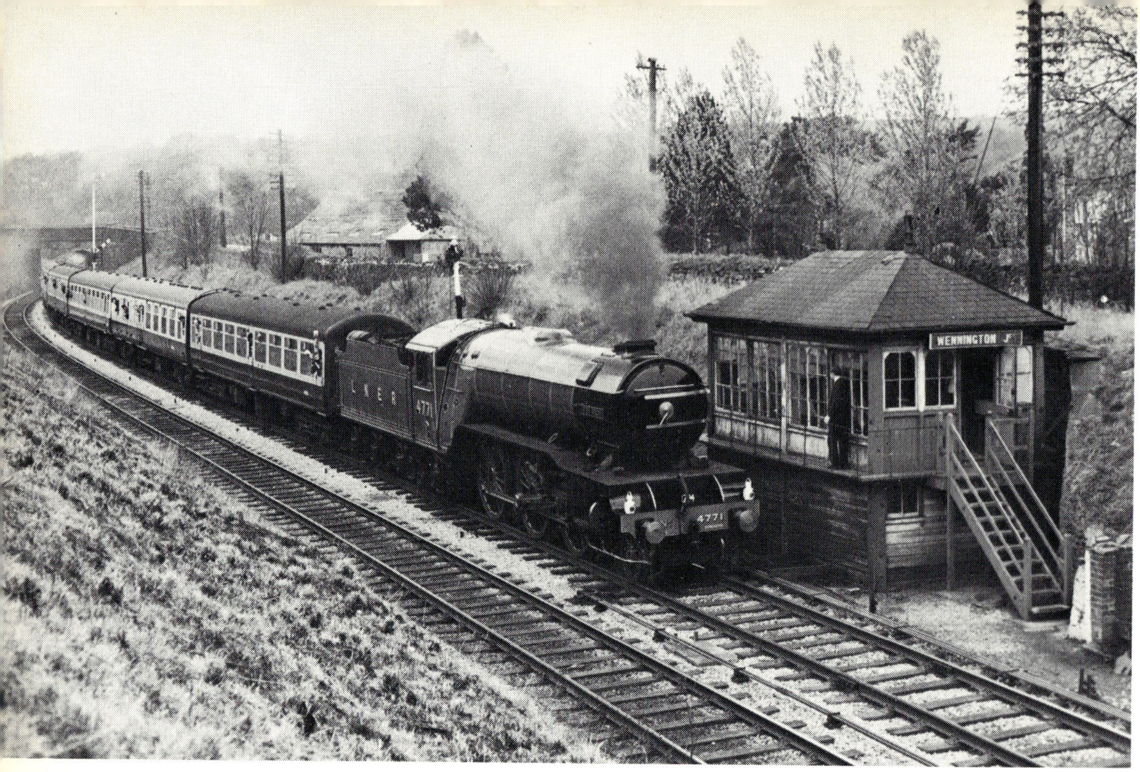
WENNINGTON
L N E R
4771
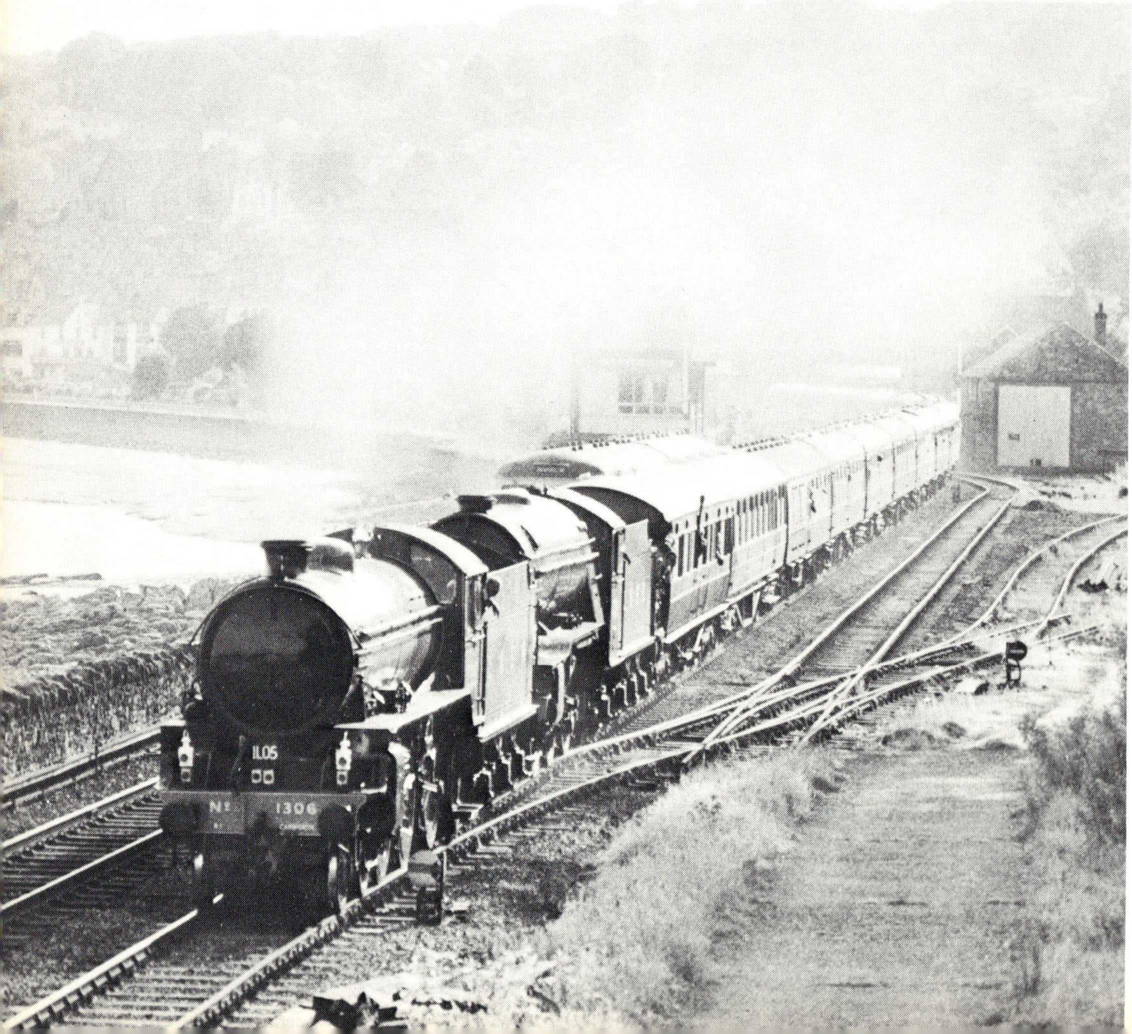
1306

Top left: LNER Class V2 No 4771 *Green Arrow* drifts past Wennington Junction signalbox with an LNER Society special from London to Leeds via Carnforth on 13 April 1974. This is a junction no more, the former Midland line to Lancaster having been lifted.
(G T Heavyside)

Above: Shades of 1895. 81 years after the Railway Race to Aberdeen LNWR 2-4-0 Precedent class No 790 *Hardwicke* leaves the Furness Railway side of Carnforth station with a train for Grange on 9 May 1976. The line to Leeds curves away in the left foreground while Steamtown shed is just out of the picture on the right.
(G T Heavyside)

Left: Green Arrow piloted by B1 class No 1306 *Mayflower* skirts Morecambe Bay at Grange-over-Sands returning from Sellafield to Carnforth on 21 June 1975. The top of a two-car dmu forming a stopping service to Barrow can be seen above the front coaches, contrasting with the heavy load behind the LNER engines.
(G T Heavyside)

Right: Reflections of *Mayflower* and *Flying Scotsman* in the Esk estuary at Eskmeals on their run round the Cumbrian coast from Sellafield to Carnforth on 8 May 1976. *(R E B Siviter)*

Below: Two weeks later the same combination leave Foxfield in more dismal Lakeland weather, after a photographic stop. The train had started at Bristol, hence the headboard 'The Merchant Venturer', a title originally carried by a Paddington—Bristol service in the palmier days of named trains on BR. *(G T Heavyside)*

Bottom right: Stanier Black 5 4-6-0s No 45407 and No 44871 running tender to tender wait to leave Ravenglass for Carnforth on 5 May 1973, patrons having visited the adjacent narrow gauge Ravenglass & Eskdale Railway. It was necessary to operate engines in this unorthodox manner, if tender first running was to be avoided, before the triangle on Ministry of Defence property at Eskmeals was made available for turning engines from 1975. Both engines can lay claim to a place in history, 44871 being one of four engines used on BR's last official steam train on 11 August 1968, while the previous week-end 45407 hauled the final steam freight. *(G T Heavyside)*

Ireland

Left: The main railways in Ireland are 5ft 3in gauge, and at the Railway Preservation Society of Ireland headquarters, Whitehead, Co Antrim, ex Londonderry Port & Harbour Commissioners No 3 0-6-0ST *R H Smyth* (Avonside No 2021 of 1928) is prepared for work on 7 August 1977. *(Charles P Friel)*

Right: Steam virtually has free rein of the Emerald Isle and on a two-day trip to the Republic GS&WR Class J15 0-6-0 No 186, only three years short of its centenary, returns along the single line branch from Youghal towards Cork on 12 June 1976. *(D Rodgers)*

Bottom right: The following day No 186 and GNR(I) Class S 4-4-0 No 171 *Slieve Gullion* thunder away up the climb from Cork towards Mallow. *(D Rodgers)*

Below: With the Portrush Flyer from Belfast York Road, a regular summer feature since 1973, LMS-NCC Class WT 2-6-4T No 4 attacks the 1 in 75 at Monkstown, heading for the Atlantic coast on 17 July 1976. *(Charles P Friel)*

1975— Rail 150 Celebrations

Left: The 150th anniversary of the Stockton & Darlington Railway was celebrated by an unprecedented gathering of preserved locomotives at British Rail Engineering Ltd, Shildon, where LBSCR rebuilt 'Terrier' A1X 0-6-0T No 72 *Fenchurch*, from the Bluebell Railway, nestles between GWR 0-6-0PT No 7752 (Birmingham Railway Museum) and LMS Class 5 4-6-0 No 4767 *George Stephenson* (North Yorkshire Moors Railway) on 29 August 1975. *(G T Heavyside)*

Below: The highlight of the celebrations was a cavalcade of 33 steam engines (plus one electric and an HST set) from Shildon to Heighington, appropriately led by the replica *Locomotion*, here passing Heighington on Sunday 31 August 1975. *(G T Heavyside)*

Right: Over 250,000 people were thought to have witnessed the cavalcade, and at Heighington a small proportion greet SR Class S15 4-6-0 No 841 *Greene King*. *(G T Heavyside)*

Bottom right: All good things come to an end! Following the cavalcade LNER Class D49 4-4-0 No 246 *Morayshire* and CR 0-4-4T No 419 leave Darlington returning to their Scottish Railway Preservation Society depot at Falkirk. *(G T Heavyside)*

On the picturesque Esk Valley route from Whitby to Battersby, a line promoted by BR in recent years, K1 class Mogul No 2005 blends well with the scenery near Lealholm on 28 June 1975. Steam was allowed on this route in 1975 as part of Rail 150 but unfortunately some trips had to be cancelled due to exceptionally dry weather and the attendant fire hazard. *(J S Whiteley)*

On 28 September 1975 with the last of the four Rail 150 specials to use the East Coast main line A4 Pacific No 4498 *Sir Nigel Gresley* leaves Sheffield, at Tinsley, on the first stage of the steam-hauled journey to Newcastle. *(J S Whiteley)*

The camera catches a glimpse of A3 Pacific No 4472 *Flying Scotsman* hurtling north, with the train seen below left, soon after passing Thirsk on the famous racing track between York and Darlington, scene of many earlier steam exploits. *(G T Heavyside)*

Passing Thirsk on the second of its three journeys to Newcastle *Flying Scotsman* is piloted by B1 class 4-6-0 No 1306 *Mayflower* on 21 September 1975. *Flying Scotsman* was a last minute replacement for *Green Arrow* at York after the latter had hauled the train from Sheffield. *(G T Heavyside)*

The Might and Majesty of Steam

The might and majesty of steam is typified by this montage of photographs....

1 GWR 4-6-0s Nos 7808 *Cookham Manor* and 6998 *Burton Agnes Hall* near Stratford-upon-Avon.
2 LNER A4 Pacific No 4498 *Sir Nigel Gresley* at York.
3 SR U 2-6-0 No 1618 on the Bluebell Railway.
4 LMS 2MT 2-6-0 No 46443 on the Severn Valley Railway.
5 BR Standard 4MT 4-6-0 No 75078 on the Keighley & Worth Valley Railway.
6 GWR 2-6-2T No 6106 at Didcot.

... long may the reign continue.
(G T Heavyside)